Make:

Easy
Electronics

Charles Platt

Maker Media, Inc.
San Francisco

Easy Electronics
Charles Platt

Maker Media books may be purchased for educational, business, or sales promotional use. Online editions are also available for most titles (safaribooksonline.com). For more information, contact our corporate/institutional sales department: 800-998-9938 or corporate@oreilly.com. Customers may purchase kits, books, and more directly from us at Maker Shed (scan code below).

Publisher: Roger Stewart
Illustration and Design: Charles Platt
Fact checkers: Jeremy Frank, Russ Sprouss, Gary White, and Marshall Magee.
November 2017: First Edition
Revision History for the First Edition
2017-11-15 First Release
See oreilly.com/catalog/errata.csp?isbn=9781680454482 for release details.

978-1-680-45448-2 [LSI]

How to Use This Book

I wrote this book to help you get acquainted with electronics more simply, more quickly, and more affordably than has ever been possible before. A dozen hands-on experiments will show you the basics, and each should take half an hour or less.

You won't need any tools. *No tools at all.*

You will need a few parts. The Shopping List on page 47 will help you to buy parts online, but there's an easier way. A small, affordable kit of components has been developed specially for *Easy Electronics.*

If your phone can read the code on the left, it will take you to a web site to buy the kit. If you have trouble with the code on the left, try the code on the right.

If you prefer, use this link:

www.protechtrader.com/easyelectronics

Problems? You can contact me directly at

make.electronics@gmail.com

—Charles Platt

Reader Registration, Feedback, and Advice

By registering your email address, you will receive any error corrections or updates relating to this book, plus plans for a "free bonus project." Just send an email with REGISTER in the subject line, to:

make.electronics@gmail.com

Your address will not be shared with any other company or individual. Once or twice a year I may send information if I publish a new book, but I won't use your address for any other purpose.

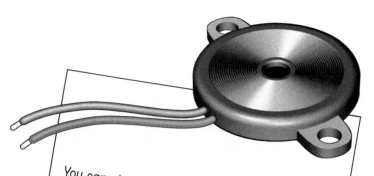

You can also go to my web site at

www.plattelectronics.com

where you will find free videos, ideas for new projects, news about interesting components, and other updates.

If you start to feel the same fascination that I feel about this subject, you can explore it through a more detailed book such as my own *Make: Electronics*. This will teach you how to use tools, ranging from a soldering iron to a multimeter, while you build circuits of increasing complexity.

Experiment 1
Power and Light

The parts on this page are all you need for the first experiment. See page 47 for a complete list of parts for all the experiments.

One **holder** for a set of three batteries.

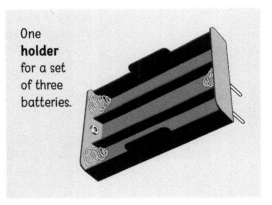

Two miniature 5-volt **light bulbs.** (Not LEDs, because a bulb is easier to use and shows how much power you are getting from a battery. I'll use LEDs later in the book.)

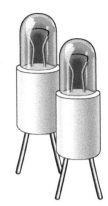

Three **alkaline batteries,** AA size. The end marked with a plus sign is **positive.** Please use fresh batteries for optimum results. (*Do not use lithium batteries.*)

Two **holders** for single batteries.

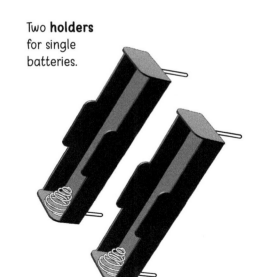

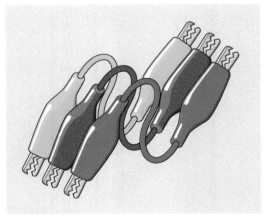

Three wires like these. The clips are called **alligator clips**, so I'm going to call the wires **alligator wires.** It's okay if your wires are longer, but they may get more tangled.

Install a battery in the holder with the negative end against the spring. The two **pins** at the back connect with positive and negative ends of the battery.

The light bulb glows more brightly when you add a second battery.

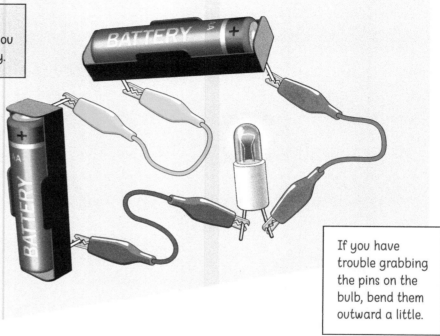

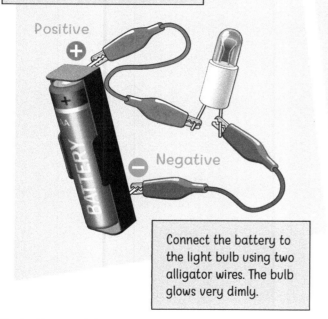

Positive ⊕

Negative ⊖

Connect the battery to the light bulb using two alligator wires. The bulb glows very dimly.

If you have trouble grabbing the pins on the bulb, bend them outward a little.

The battery and the light bulb are **components.** When you connect them, you create a **circuit.** Imagine electricity flowing from the positive end of the battery, through the red wire, through the bulb, and back through the black wire to the negative end of the battery.

Electricity consists of tiny particles in the wires, known as **electrons.** They have a negative charge, but when Benjamin Franklin was experimenting with electricity (before he got his picture on the $100 bill) he decided that electricity flows from positive to negative, and we still think of it that way.

The flow of electrons during a period of time is called **current**.

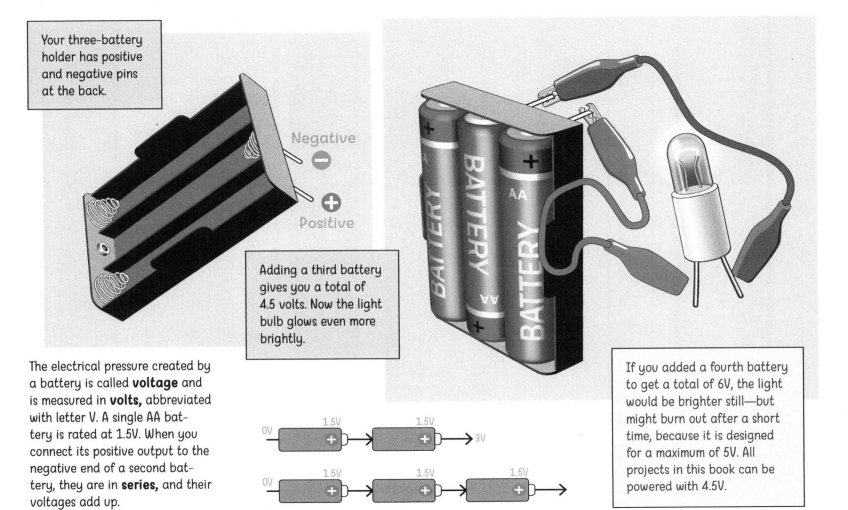

Your three-battery holder has positive and negative pins at the back.

Negative

Positive

Adding a third battery gives you a total of 4.5 volts. Now the light bulb glows even more brightly.

The electrical pressure created by a battery is called **voltage** and is measured in **volts,** abbreviated with letter V. A single AA battery is rated at 1.5V. When you connect its positive output to the negative end of a second battery, they are in **series,** and their voltages add up.

0V   1.5V   1.5V   3V

0V   1.5V   1.5V   1.5V

If you added a fourth battery to get a total of 6V, the light would be brighter still—but might burn out after a short time, because it is designed for a maximum of 5V. All projects in this book can be powered with 4.5V.

7

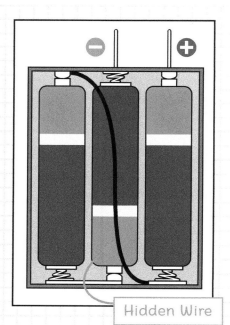

Hidden Wire

The batteries in the holder don't look as if they are in series, but they are, because the holder contains a hidden wire. Electricity doesn't care if it follows a zig-zag path.

What happens if you make a circuit where two or three batteries are beside each other? Now the batteries are in **parallel,** and their voltages don't add up anymore. You can connect any number of them in parallel, and you'll still get just 1.5V.

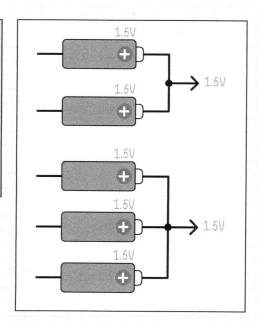

1.5V
1.5V
1.5V
1.5V
1.5V
1.5V
1.5V

Two batteries in parallel should last twice as long as a single battery, because they are sharing the work.

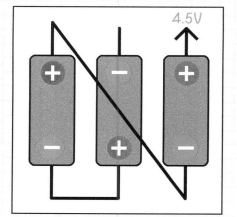

4.5V

Try adding another light bulb beside the first. Now the **bulbs** are in **parallel**.

Each of the two bulbs in parallel will be as bright as when you had just one.

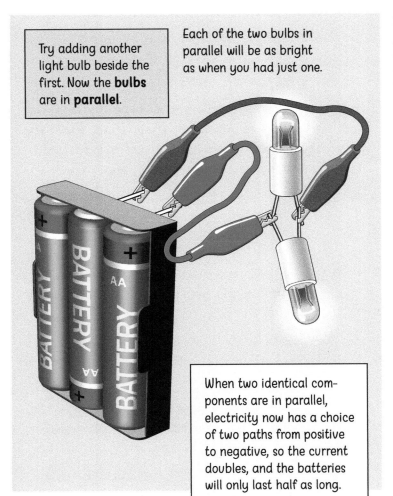

When two identical components are in parallel, electricity now has a choice of two paths from positive to negative, so the current doubles, and the batteries will only last half as long.

Move the light bulbs so that one follows the other. Hold them together with your finger and thumb. The bulbs are now in **series**.

Each of the bulbs in series will be less bright than before. The first will take half the voltage, and the second will take the other half.

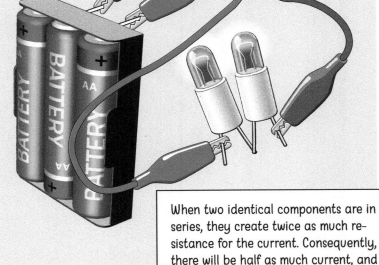

When two identical components are in series, they create twice as much resistance for the current. Consequently, there will be half as much current, and the batteries will last twice as long.

9

How Did It Work?

When you connected power to the light bulb, it took just a moment to react. It was warming up. This kind of light contains a thin piece of wire, called a **filament,** which is heated by electricity flowing through it. The heat makes it glow.

Any object that gives off light as a result of heat can be described as **incandescent.** So, you have been playing with an incandescent light bulb.

Incandescent lights are not used so often in houses and offices anymore. We use **fluorescent** lamps or **LEDs.** I'll get to LEDs later.

Amperes

Current is measured in **amperes**, abbreviated as **amps**. It is represented with letter **A**. Small currents are measured in **milliamperes**, abbreviated as **milliamps**, represented with **mA**. There are 1,000 milliamps in 1 amp (that is, 1,000mA = 1A).

You can think of volts as measuring the pressure that forces electrons into a wire, while the flow of electrons per second through the wire is measured in amps.

Schematics

A diagram that shows how components are connected is called a **schematic.** Here are the schematic symbols for an incandescent bulb and two batteries.

Light Bulb

Battery

Bigger Battery

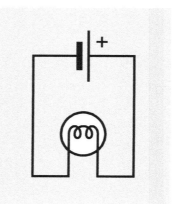

Circuit

This is a schematic version of the first circuit that you made. Can you draw schematics of all the circuits that you built so far?

After you finish each experiment in this book, unclip the red wire from the battery to break the circuit. Otherwise, the battery will run down.

Switching

A switch can turn power on and off, or it can select different parts of a circuit.

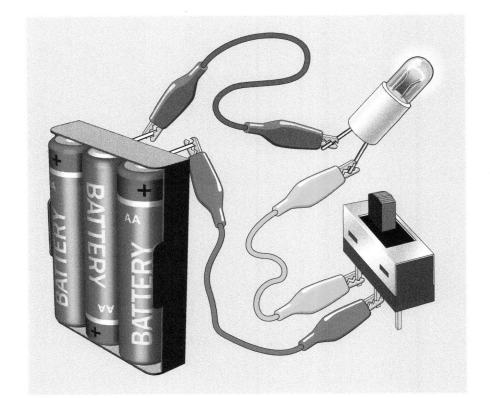

There are many thousands of different kinds of switches. Search online for "types of switch," and look at all the images you find.

All switches have some kind of actuator. It may be a lever that pivots, a knob that turns, or a button that slides.

All switches have terminals. They may be screws, lugs, tags, wires, or pins.

This is a **slide switch.** The plastic button is called an **actuator** and has two positions (some switches have more). The metal pieces sticking out at the bottom are **terminals.** Terminals that look like this are called **pins.**

The terminal in the middle of this switch makes a connection with one of the terminals at each end, depending which way you move the actuator. A switch that has two positions like this is called a **double-throw** switch.

A double-throw switch can be used as an on-off switch if you don't attach anything to one of its terminals.

11

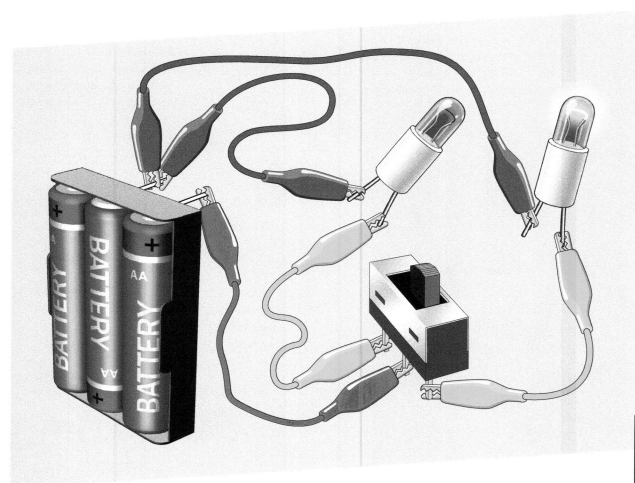

You can use a double-throw switch to choose between two light bulbs. If you switch the positive wire, you can have more than one negative wire sharing the negative side of the power supply.

Transistors can be used as switches. You'll see how in Experiment 4.

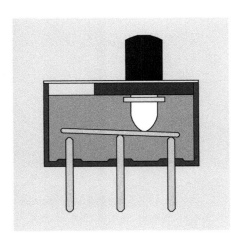

How Did It Work?

Inside this double-throw switch are three **contacts.** The center contact is called the **pole.**

A pointed piece of white nylon is pressed downward onto a brass strip by a tiny hidden spring. When the actuator moves left or right, the nylon makes the strip pivot around the pole to connect with one contact or the other.

Schematics

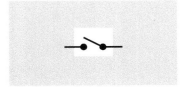

If a switch has two contacts, it is a **single-throw** switch, also called an **on-off** switch.

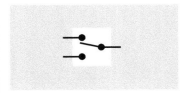

This is the symbol for a generic **double-throw** switch.

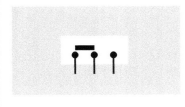

A **slide switch** may be shown like this, or the generic symbol can be used.

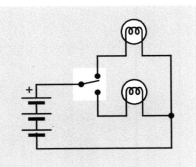

This is a schematic version of the circuit shown on the previous page. A **dot** shows where two wires are connected to each other.

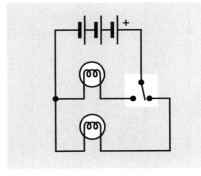

This is the same circuit as above. The parts have been moved around, but they are still connected the same way, so the circuit will still work the same way.

You can add an on-off switch to any circuits that you build, so you don't have to unclip alligator wires to disconnect power.

Experiment 3
Revealing Resistance

Put the resistor into series with your light bulb. Touch the resistor lead to the pin on the light.

A resistor is a little component that usually has stripes printed on it. All it does is resist the flow of electricity—but that can be very useful. See page 48 if you want to know which resistors you will need for this experiment and the ones that follow.

The colored stripes are a code telling you the resistance of the resistor. I will explain the code in a moment. For this experiment, you need a "33-ohm" resistor with stripes that are orange, orange, and black.

Ignore the silver or gold stripe at the opposite end.

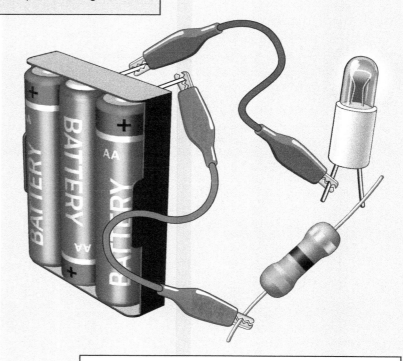

The wires at each end are called **leads** (pronounced "leeds").

The resistor limits current and drops voltage, leaving less available for the light, which glows less brightly than if you touch the red alligator clip to it directly.

Try putting two resistors in series. The light is even dimmer, now. Electric current has to push through two resistors before it gets to the light.

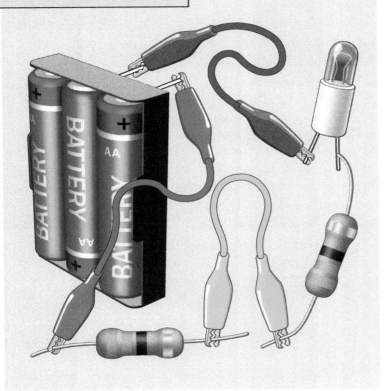

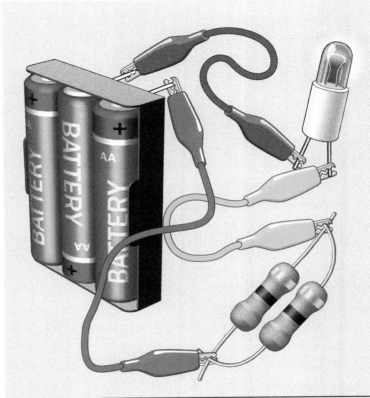

Try putting the resistors in parallel. Now the current can flow through both of them, so the light gets brighter—although not as bright as with no resistors at all.

Understanding the Resistor Code

Each of the first two colored stripes on a resistor tells you a single digit. The third stripe tells you how many zeroes to add.

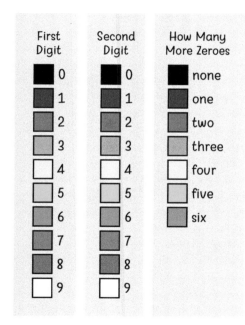

First Digit	Second Digit	How Many More Zeroes
0	0	none
1	1	one
2	2	two
3	3	three
4	4	four
5	5	five
6	6	six
7	7	
8	8	
9	9	

A silver stripe at the right end of the resistor means that its value is accurate within 10%. A gold stripe means 5%. Either will be okay in this book.

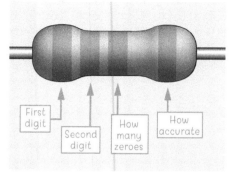

First digit
Second digit
How many zeroes
How accurate

Hold the resistor with its group of three stripes on the left.

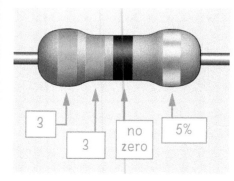

3
3
no zero
5%

A resistor with orange-orange-black stripes has a value of 33 ohms.

Resistance is measured in **ohms.**

Capital Letter K means 1,000 ohms, so 2K is 2,000 ohms, 3.3K is 3,300 ohms, and 470K is 470,000 ohms.

Capital Letter M means 1,000,000 ohms. So 2M is 2,000,000 ohms, and 1.5M is 1,500,000 ohms.

Sample Resistor Values

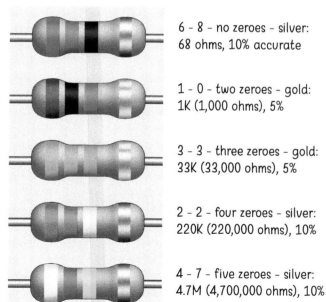

6 - 8 - no zeroes - silver: 68 ohms, 10% accurate

1 - 0 - two zeroes - gold: 1K (1,000 ohms), 5%

3 - 3 - three zeroes - gold: 33K (33,000 ohms), 5%

2 - 2 - four zeroes - silver: 220K (220,000 ohms), 10%

4 - 7 - five zeroes - silver: 4.7M (4,700,000 ohms), 10%

How Did It Work?

Most resistors contain carbon film or metal film that has a resistance higher than copper. The body of the resistor is covered in paint or plastic that is usually beige, although its color doesn't matter. Just pay attention to the colors of the stripes.

If you're wondering what happens to the current blocked by the resistor, it is converted to heat. There isn't enough for you to feel it with your fingers, because these circuits use so little power.

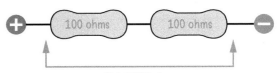

Total 200 ohms

Two 100-ohm resistors in series block twice as much current as a single resistor. Their total resistance is 200 ohms.

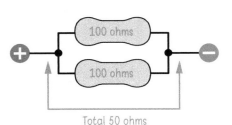

Total 50 ohms

Two 100-ohm resistors in parallel block half as much current as one resistor. Their total resistance is 50 ohms.

Schematics

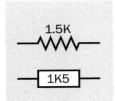

Symbols for a 1.5K resistor, in the American style (top) and European style (bottom).

Europeans don't use a decimal point in schematics. If you see 1K5 it means 1.5K, while 4M7 means 4.7M. Values less than 1,000 ohms use letter R, so 33 ohms would be written as 33R.

Schematic for a circuit with two resistors in series.

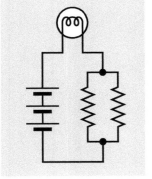

Schematic for a circuit with two resistors in parallel.

Could you redraw these schematics including an on-off switch? Could you add a double-throw switch that allows you to send current through a resistor or bypass it through a wire?

Experiment 4

Try a Transistor

A transistor contains no moving parts, but can work like a switch.

The experiments in this book use a transistor whose part number is 2N3904. This is a very cheap, widely used NPN bipolar transistor. I won't delve into the technical terms here, but you can easily find them online. I am mainly interested in showing you what a transistor does.

Some versions are packaged in tiny metal cans, but most look like this.

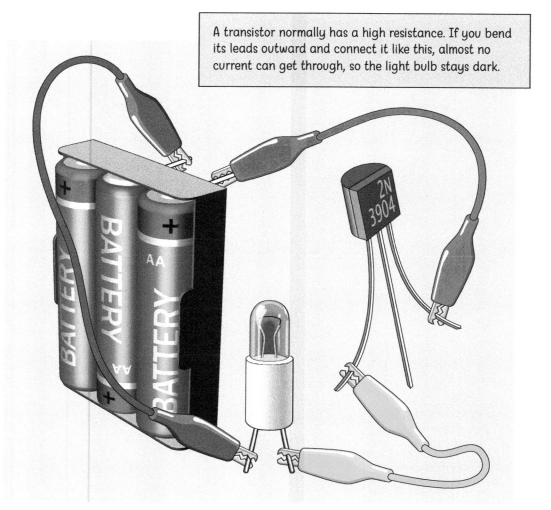

A transistor normally has a high resistance. If you bend its leads outward and connect it like this, almost no current can get through, so the light bulb stays dark.

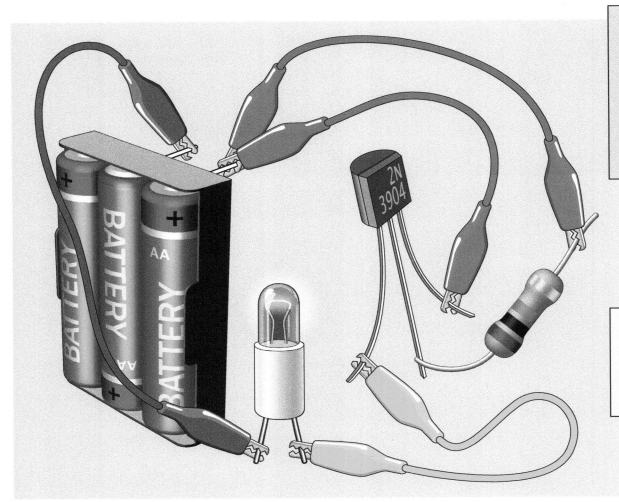

If you apply positive voltage to the lead in the middle, this tells the transistor to allow current to flow, and the light comes on. Use a 1K resistor (with brown, black, and red stripes) to protect the center lead from too much voltage. It is very sensitive.

Instead of the 1K resistor, try a 10K resistor (with stripes that are brown, black, orange). This reduces the current to the center lead, and the light is much dimmer.

The transistor is **amplifying** the current supplied to its center lead.

How Did It Work?

A transistor is a **semiconductor.** It is not quite a conductor because its resistance varies.

In this experiment you used a **bipolar junction transistor**. Its name refers to the junctions inside it between two types of silicon. N-type silicon has more electrons, while P-type has fewer, because the silicon has been specially treated.

The layers are called the **collector, base,** and **emitter.** In an **NPN** transistor, a small positive current flowing into the base will allow a larger current to flow into the collector and out of the emitter. The resistance between the collector and the emitter decreases as the base current increases, until the resistance is almost zero. At this point, the transistor is **saturated**.

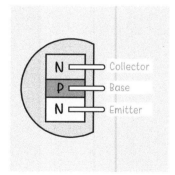

If you connect a transistor the wrong way around, it will pass some current, but can easily be damaged. Many electronic parts are sensitive about being connected incorrectly. We say they have **polarity.** If you see a warning to "observe polarity," check a book or an online source to figure out which side of the component has to be more positive than the other.

Schematics

This is the symbol for an NPN bipolar transistor.

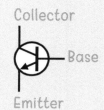

In an NPN transistor, the collector should be more positive than the base, and the base should be more positive than the emitter.

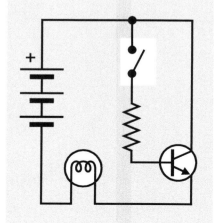

Here's the schematic for the circuit in this experiment. I added an on-off switch to show that you can decide whether to apply current to the base.

I turned the transistor around because it fits more easily that way. Any component can be flipped in a schematic. Its function remains unchanged.

Experiment 5

Lighting an LED

Your skin has electrical resistance, just like a resistor. Try using your finger instead of the resistor from the previous experiment. Can you make the light come on?

Don't let the alligator clip make contact with the center lead, without your finger between them. That can overload the transistor.

It's okay to touch bare wires in the experiments in this book, because the voltages and currents are so low. But don't get in the habit of sticking your fingers in other electronic devices, which may contain high voltages. People are killed by electricity each year. Please don't become one of them.

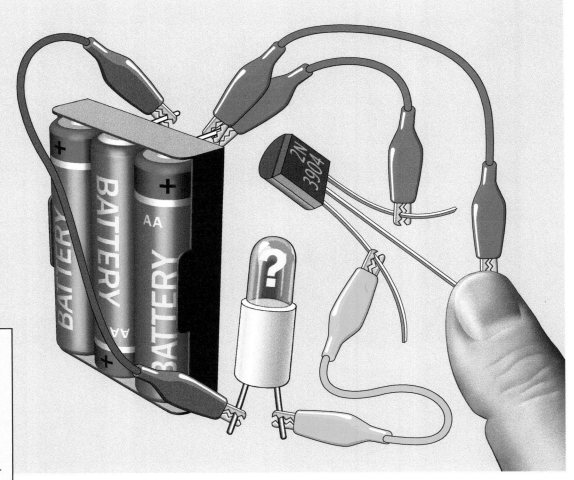

If you use an LED instead of the little incandescent bulb, you'll get better results. An LED is more sensitive and needs less power. Use the efficient type of LED recommended for this experiment in the Shopping List on page 47.

Press hard with your finger, and the LED may glow dimly. Wet your finger and try again, and it will glow more brightly, because electricity flows more easily through the moisture.

Make sure the LED is the right way around when you connect it to the batteries. One lead is longer than the other, and must be more positive than the other. This is known as **polarity**. If the LED is the wrong way around, you can burn it out.

There are thousands of types of LEDs, including some that are used for room lighting. The one in this experiment is the old original "standard" LED, 5mm in diameter, used as an indicator in electronic gadgets.

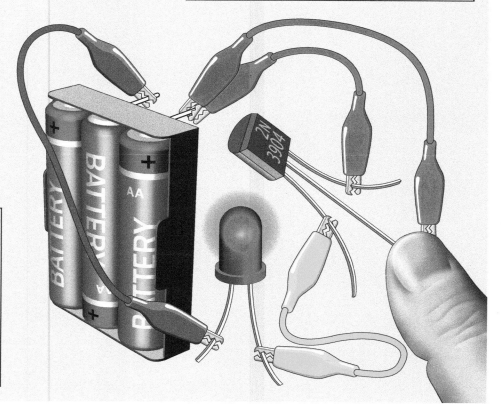

How Did It Work?

LED is an acronym for **light-emitting diode.** A diode is a semiconductor that contains layers of silicon, like a transistor, except there are only two instead of three. In an LED, they have been specially made to convert electrons into **photons,** which are particles of light.

The LED that I have specified only needs 1mA of current at 1.6V. The incandescent light bulb that I recommended would like to have 60mA at 5V. That's almost 200 times as much power!

An incandescent light bulb wastes a lot of power as heat. The LED uses much more of its power to create light.

I wanted you to use a light bulb in the first four experiments because it has no polarity to worry about. You can connect it directly to a battery, and it responds to a wide range of voltage. The brightness of the light indicated the voltage that it was receiving.

You'll notice that the LED is brighter when your skin is wetter, when the contacts are closer together, or when you press your finger harder. These are three ways to reduce the resistance in the circuit.

> Never connect an LED directly to a battery, as you did with an incandescent light bulb. You need a resistor to stop too much current from damaging the LED.

Schematics

This is the symbol for an LED. The big triangular arrow points in the direction that current must flow. The small arrows tell you that the component emits light.

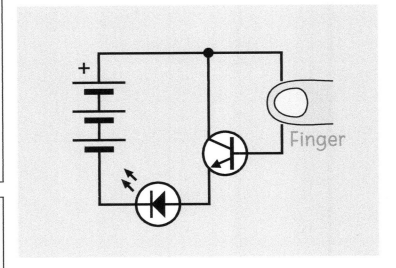

Here's the schematic for the circuit. There's no symbol for a finger, so I just made one up.

Experiment 6

Seeing Light

A phototransistor has **polarity**. In this case, the long lead must be more positive than the short lead. (Some phototransistors are the other way around.)

Use the same 1K resistor that you used before (brown, black, red).

A **phototransistor** works like a regular transistor, except that it has no electrical connection to its base. The base is controlled by light instead of electricity.

You're going to see how the phototransistor can turn an LED on and off.

Many phototransistors look like LEDs. That can be confusing, so I chose one that looks as different from an LED as possible. It's a **side-facing** phototransistor. The bump is a lens that must point at a light source.

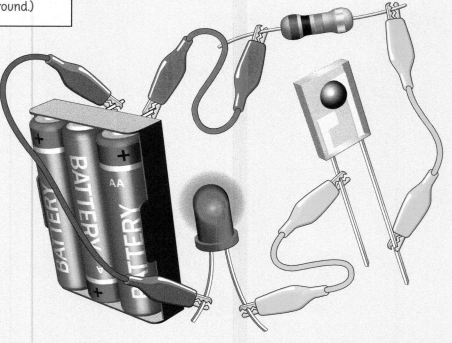

Cover the phototransistor to shield it from light. The LED remains dark. Shine a strong light on the phototransistor, and the LED comes on.

How Did It Work?

A phototransistor is a semiconductor, so its resistance varies.

Light falling on the phototransistor forces its resistance low. This allows enough current to light the LED.

When you protect the phototransistor from light, its resistance rises very high. This cuts off the current.

The 1K resistor is included to limit the amount of current flowing through the phototransistor and the LED. Neither of them is designed to pass a lot of current.

This circuit requires the same low-power LED that I recommended for the previous experiment.

If necessary, point a brighter light at the phototransistor. A laser pen or laser cat toy is ideal. You can also try using a 2.2K resistor (red, red, red) or a 3.3K resistor (orange, orange, red) instead of the 1K resistor (brown, black, red).

Concept for a Project

Suppose you mount the phototransistor in a little cardboard tube, to protect it from light, but you put a laser pen about three feet away, pointing directly into the end of the tube.

If someone walks through the laser beam, the phototransistor sends a signal down a long wire to sound an alarm of some kind. This is a very simple intrusion alarm, which I will develop in Experiment 12.

Schematics

The symbol for an NPN phototransistor includes 2 arrows pointing inward, to remind you that it responds to incoming light. The lead at the top must be more positive than the lead at the bottom.

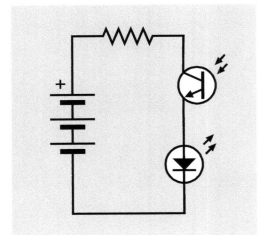

The phototransistor must be protected from high current. The resistor takes care of this.

Some LEDs would overload the phototransistor in this circuit. You need the kind recommended in the Shopping List that begins on page 47.

Experiment 7

Comprehending Capacitors

A capacitor stores electricity, as you will see in this experiment.

This type of capacitor is called **electrolytic.** Its storage capacity, known as **capacitance,** is 470μF—but I'll explain that in a moment. 50V is its maximum voltage, but for the experiments in this book, a capacitor rated 10V or higher is okay.

The short lead is the negative side, also identified with minus signs. Never connect an electrolytic capacitor to a power supply the wrong way around.

Some capacitors have colored cans. Others don't. The color is not important.

This is a larger circuit, so you can build it in two parts. This part just charges the capacitor with electricity when the slide switch moves to upper-left.

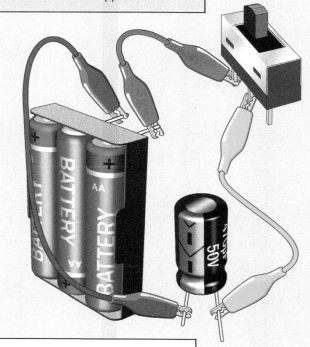

Some of the voltage from the battery transfers to the capacitor, although you can't see any sign of it yet.

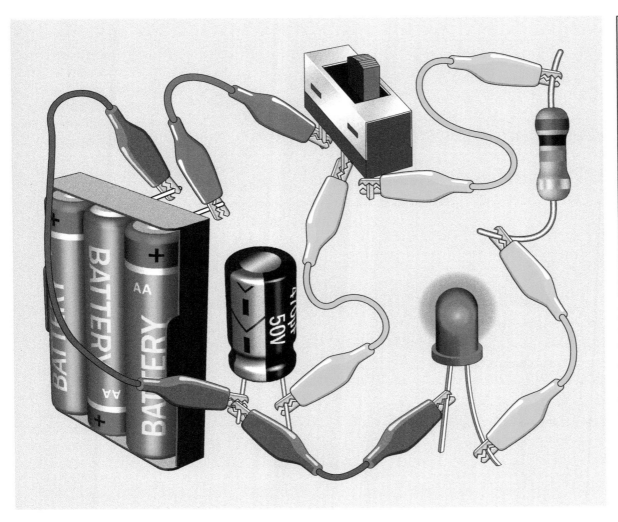

Add the 1K resistor and the LED, with the negative side of the LED sharing the negative leg of the capacitor.

Now move the switch to the lower-right. The capacitor discharges itself through the LED.

Move the switch to the upper-left and wait five seconds for the capacitor to recharge. Now you can discharge it again.

If this diagram looks complicated to you, try sketching a copy of it, replacing the alligator wires with simple lines to connect the components. Or check the schematic on page 29.

Ceramics

Capacitors such as the one on the right are less than 1/2" wide. They are dipped in a **ceramic** compound.

Most ceramic capacitors do not have polarity.

Many ceramic capacitors have a code printed on them instead of their actual capacitance.

Some ceramic capacitors are shaped like circular discs.

In simple circuits of the type you have been building, usually you can substitute a ceramic instead of an electrolytic if you wish. Note that for values around 10µF and above, ceramics may be more expensive.

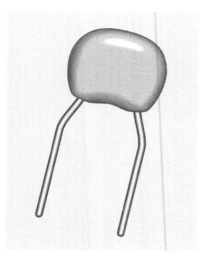

Units

Capacitance is measured in farads, abbreviated with letter F. But a 1F capacitor is very large. In hobby electronics we mostly use capacitors rated in microfarads, abbreviated µF. The µ symbol is the Greek letter mu, but often µF is printed as uF.

There are 1,000,000 microfarads in 1 farad, 1,000 nanofarads (nF) in 1 microfarad, and 1,000 picofarads (pF) in 1 nanofarad.

How Did It Work?

Inside the capacitor you used are two pieces of metal film known as **plates.** They are separated by paste called an **electrolyte,** which is why this capacitor is called electrolytic.

When electrons flow into one plate, they try to create an equal, opposite charge on the other plate. You can think of the plates as having positive and negative charges that attract each other.

Timing

The 1K resistor was needed because you charged the capacitor with 4.5V from the battery pack, and the LED can only handle about 1.8V. The resistor prevents the LED from being damaged.

The resistor also controls how fast the capacitor discharges. Substitute a 10K resistor (brown, black, orange) and the LED is dimmer than before and takes much longer to fade out.

Here's another thing to try. Go back to using the 1K resistor. Remove the 470µF capacitor and substitute a 100µF capacitor. Push the switch to and fro, and now the LED lights up very briefly.

Electricity moves fast, but a capacitor and a resistor can make things happen slowly.

Schematics

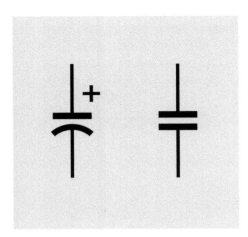

A capacitor may seem similar to a battery. After all, they both store electricity.

A battery, however, uses chemical reactions, and even a rechargeable battery wears out after a limited number of charging and discharging cycles.

A capacitor does not use chemical reactions, and can still work as well after several years.

There are two symbols for capacitors.

A **polarized** capacitor, such as an electrolytic, is on the left.

A **nonpolarized** capacitor, such as a ceramic, is on the right.

Some people use the symbol on the right everywhere in a schematic, and let you decide if you want to use an electrolytic capacitor, and if so, which way around it should be.

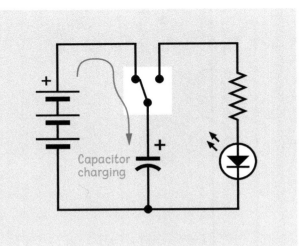

In this schematic showing the circuit that you just built, the double-throw switch has completed a circuit with the battery, so that the battery charges the capacitor.

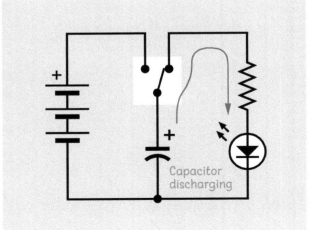

In this schematic, the double-throw switch is in its other position, completing a circuit from one plate of the capacitor, through the resistor and the LED, back to the other plate.

A Simple Chip

The rest of the experiments in this book will use an **integrated circuit chip.** It contains tiny transistors on a little piece of silicon, sealed inside a plastic package.

This chip is called a **7555 timer,** although yours will have extra letters and numerals printed on it.

This experiment will test the chip by flashing an LED.

The legs of the chip are too closely spaced for you to grab them with alligator clips, so you will plug it into a **breadboard.** (Why is it called that? Because many decades ago, test circuits were built on real wooden boards.)

This is a mini-breadboard. Hidden inside it are metal strips linking the holes in groups of five. The strips connect the leads of components that you push into the holes.

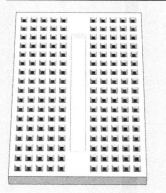
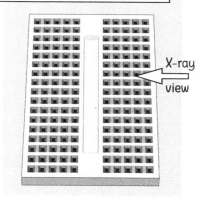

X-ray view

All chips have a half-circle or (sometimes) a dimple molded into one end. Always keep it facing the top of the breadboard. Plug in the chip, leaving 6 rows empty above it. Count carefully!

To make extra connections, you'll use **jumper wires**. They are included in the kit for this book. You can also buy a set online, or you make your own with 22-gauge hookup wire and a wire stripper (search YouTube for a demo).

If leads on components are too long, trim them with nail clippers, cutting at an angle to create sharp points. Use eye protection if you are concerned about flying fragments.

1. Add a 100K resistor (brown, black, yellow) beside the chip, in exactly this position. Count the holes carefully!

2. Add a jumper wire. Use green jumpers when you are not making connections with a power supply. The color helps to remind you of the function of each wire.

3. Add a 10K resistor (brown, black, orange). All the components must be in exactly the right locations.

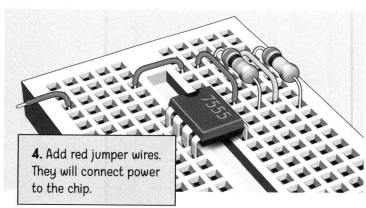

4. Add red jumper wires. They will connect power to the chip.

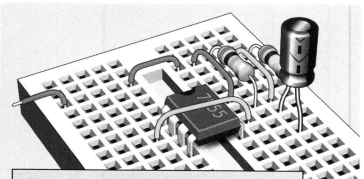

5. Add a 10µF capacitor with the minus signs and the short lead facing the bottom of the board. Add a jumper wire across the chip. Be careful! The left end of the wire has 7 empty rows above it, but the right end has 8 rows above it.

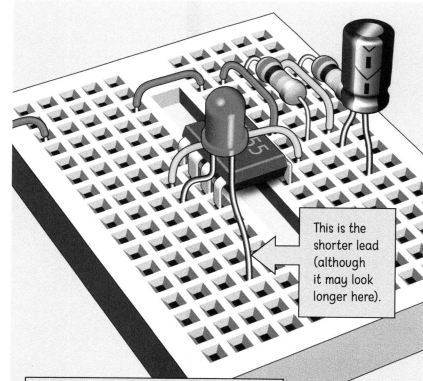

This is the shorter lead (although it may look longer here).

6. Now add an LED. The *shorter* lead must face the *bottom* of the board. The LED does not need a resistor to protect it from too much voltage, because the output from the chip will be below 1.8V.

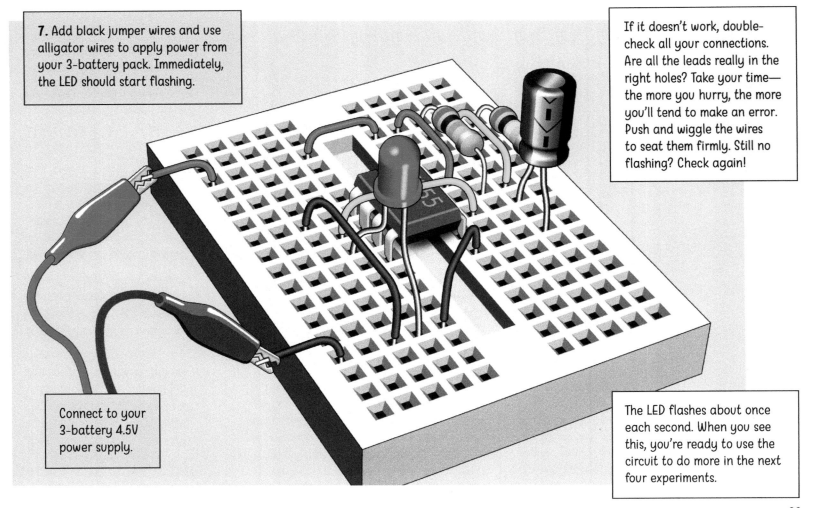

7. Add black jumper wires and use alligator wires to apply power from your 3-battery pack. Immediately, the LED should start flashing.

If it doesn't work, double-check all your connections. Are all the leads really in the right holes? Take your time—the more you hurry, the more you'll tend to make an error. Push and wiggle the wires to seat them firmly. Still no flashing? Check again!

Connect to your 3-battery 4.5V power supply.

The LED flashes about once each second. When you see this, you're ready to use the circuit to do more in the next four experiments.

How Did It Work?

The resistors and the capacitor control the speed of the pulses. The capacitor charges through both resistors, then discharges through the lower one into the chip.

The LED flashes 0.69 times per second with the 10μF capacitor. Try a 1μF capacitor instead, and the LED flashes ten times faster.

The timer is too complicated for me to explain in detail, but it is not too complicated for you to understand. You can read more about it online or in an introductory book.

The 7555 timer is actually a more recent version of an older chip, the 555 timer. The functions are the same, but it uses less power.

Search for **555 timer** online and you will find a lot of useful information.

How Fast Will It Flash?

The rate of flashing is measured in Hertz, abbreviated Hz. The table below shows alternate values for the 100K resistor. The flashing rates assume that you don't change the 10K resistor located closer to the chip.

7555 Timer	Resistor Values		
	10K	100K	1M
10μF	4.8	0.69	0.072
4.7μF	10	1.5	0.15
2.2μF	22	3.1	0.33
1.0μF	48	6.9	0.72
0.47μF	100	15	1.5
0.22μF	220	31	3.3
0.1μF	480	69	7.2
47nF	1,000	150	15
22nF	2,200	310	33
10nF	4,800	690	72
4.7nF	10,000	1,500	150
2.2nF	22,000	3,100	330
1nF	48,000	6,900	720

(leftmost column label, vertical: Capacitor Values)

Number of pulses per second (Hz) using a 10K resistor beside the chip.

Pin Numbering

Chips that look like this have a half-circle or a dimple molded into the plastic at one end. The pins are numbered counter-clockwise from that end, when seen from above.

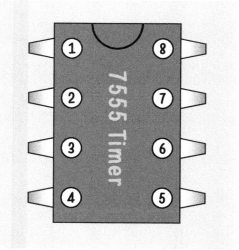

In a 7555 timer or a 555 timer, pin 1 receives negative power, pin 8 receives positive power, and pin 3 sends out a stream of pulses. Pin 4 is the reset pin, which you will use in Experiment 10.

Schematics

Compare this schematic with the breadboard layout, and you'll see that the parts are in the same positions, although the jumper from pin 2 to pin 6 goes around the chip instead of across it.

The symbol for an integrated circuit chip is just a rectangle containing the part number of the chip and the pin numbers.

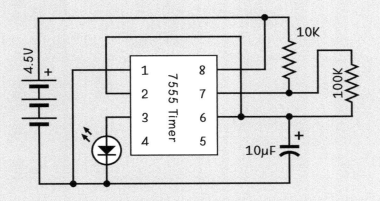

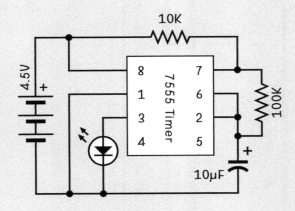

Where wires cross each other, they do not make a connection unless there is a dot joining them.

In most schematics, positive power is at the top. You can imagine current flowing from top to bottom.

Often people draw schematics with pin numbers of chips shuffled around, as in this example. But the connections between pins and components are still the same. You can avoid having wires crossing each other by shuffling the pin numbers, and the schematic can take less space. But it may be harder to understand.

Experiment 9

Sound From a Transducer

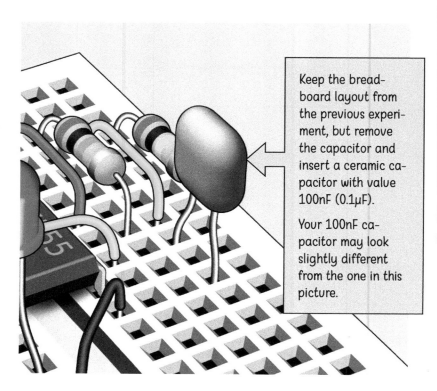

Keep the bread-board layout from the previous experiment, but remove the capacitor and insert a ceramic capacitor with value 100nF (0.1µF).

Your 100nF capacitor may look slightly different from the one in this picture.

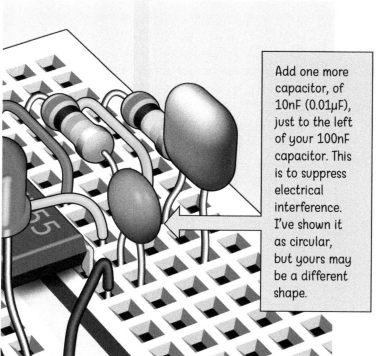

Add one more capacitor, of 10nF (0.01µF), just to the left of your 100nF capacitor. This is to suppress electrical interference. I've shown it as circular, but yours may be a different shape.

The LED now seems to be on all the time. Check the table of flashing rates on page 34, and you'll find it is flashing at 69Hz, which is too fast for you to see. But it's not too fast for you to hear!

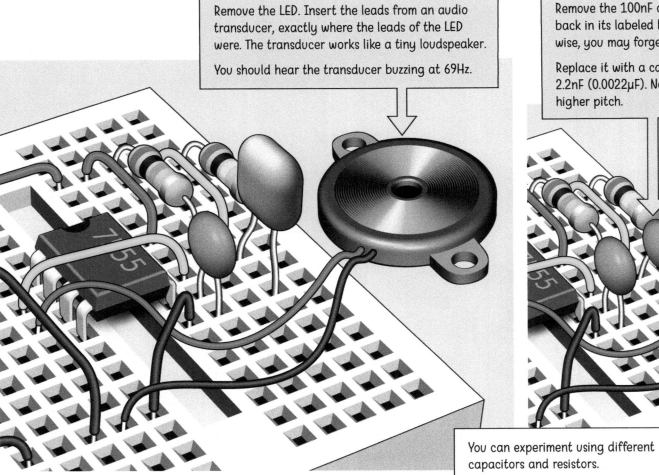

Remove the LED. Insert the leads from an audio transducer, exactly where the leads of the LED were. The transducer works like a tiny loudspeaker.

You should hear the transducer buzzing at 69Hz.

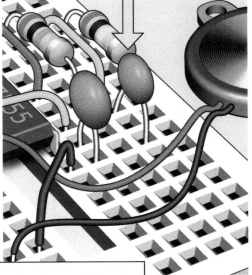

Remove the 100nF capacitor, and put it back in its labeled bag, because otherwise, you may forget what its value is.

Replace it with a capacitor of value 2.2nF (0.0022µF). Now you hear a much higher pitch.

You can experiment using different capacitors and resistors.

How Did It Work?

Sound consists of rapid changes of pressure in the air. Higher-pitched sounds consist of pulses at a faster rate.

The pulses from a timer chip move a thin wafer inside your transducer. The wafer vibrates to create pressure waves.

Pulses from the chip between 50Hz and 10,000Hz can be heard by most people, although a transducer may not reproduce some of them very well.

For a louder sound, you could double the voltage with a 9V battery. You could also amplify the output from the chip with a transistor, feeding it to a real speaker.

Multiple timer chips can be combined to create more complex sounds. You need to read more about timers to find out how.

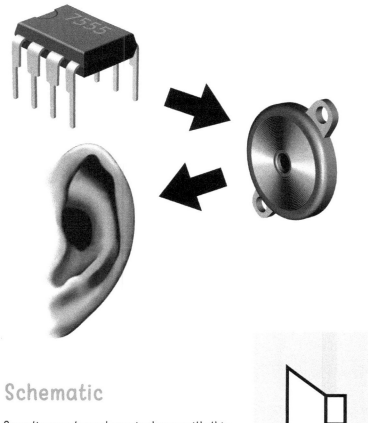

Schematic

Sometimes a transducer is shown with this symbol which is also used to represent a speaker, such as the ones in a stereo system.

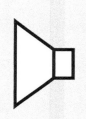

You can use the resistance of your skin instead of the 100K resistor. Remove that resistor and insert two jumper wires instead. The bare ends of the wires should stick up where you can grab them.

Move them close together, but not touching. Squeeze them between your finger and thumb. The harder you squeeze them, the higher the pitch of the sound. Wet your fingers slightly if necessary.

Put the 100K resistor back in the board before you go on to the next experiment.

Experiment 10

Touch Control

Now the pressure of your finger will start and stop the 7555 timer chip.

Keep the circuit the same as in the previous experiment, but add a red wire and a green wire sticking out at the top of the board, where you can touch the bare ends of them easily.

Add the same 2N3904 transistor that you used previously, with its rounded back facing left.

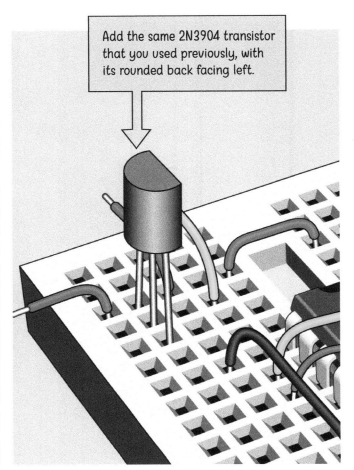

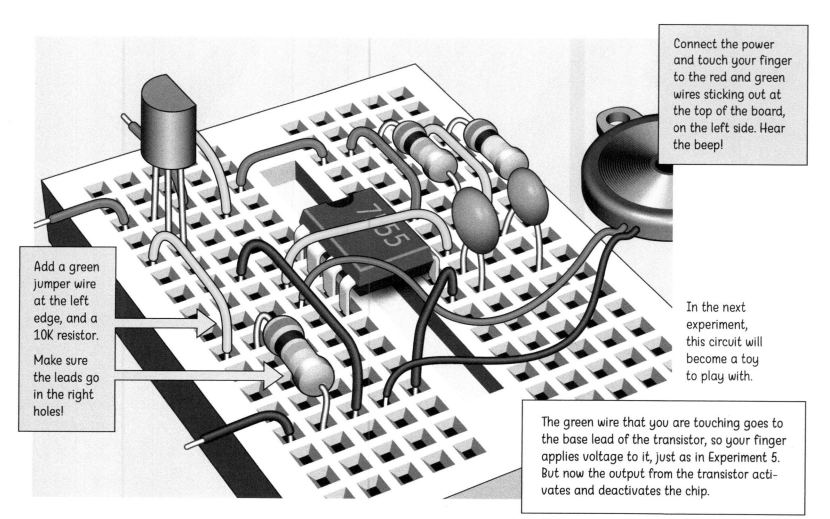

Connect the power and touch your finger to the red and green wires sticking out at the top of the board, on the left side. Hear the beep!

Add a green jumper wire at the left edge, and a 10K resistor.

Make sure the leads go in the right holes!

In the next experiment, this circuit will become a toy to play with.

The green wire that you are touching goes to the base lead of the transistor, so your finger applies voltage to it, just as in Experiment 5. But now the output from the transistor activates and deactivates the chip.

How Did It Work?

The first two schematics on the right show how the components that you added are connected with the timer chip. Compare them with the layout on your breadboard.

The output from the emitter of the transistor goes to pin 4, which is called the **reset** pin. This pin is also connected through a 10K resistor to the negative side of the power supply.

Low voltage on pin 4 makes the chip go into **reset mode,** which means that it stops its output and does nothing. **Positive** voltage keeps the chip functioning **normally.**

When you are **not** touching the jumper wires, the effective resistance of the transistor is very high compared with the 10K resistor. So the chip sees the negative voltage through the resistor and goes into reset mode.

When you touch the wires, your finger conducts electricity to the base of the transistor, so its effective resistance goes low compared with the 10K resistor. Consequently the voltage to pin 4 of the chip goes up, and the chip starts creating pulses that make noise.

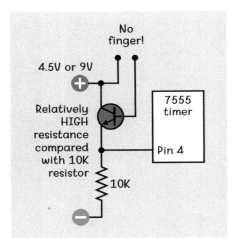

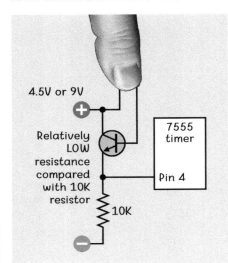

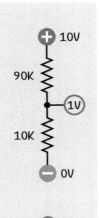

Suppose you put two resistors in series between the positive and negative sides of a 10V power supply. The voltage you measure between the resistors depends on their values. This is known as a **voltage divider**. The circuit that you just built works on this principle.

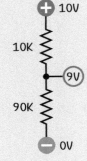

An LED is sensitive to current. A transistor amplifies current—so it was a good choice to run the LED in Experiment 5. But the timer chip is sensitive to voltage, not current. It needed a voltage divider to give it what it wants.

Your chip still uses a little power in reset mode, so don't leave it connected to the batteries, even when the transducer is silent.

Experiment 11

A Beeping Box

Here's a prank you can try on a friend. Hand them a box and say, "I've captured a mouse inside!" When their fingers touch the box, a transducer inside it starts squeaking (maybe a *little* bit like a mouse).

You will also need about 6" from a roll of aluminum foil, some Scotch tape, and scissors. Heavy aluminum foil is easiest to work with.

Cut two pieces of foil about 2" x 2". Fold them over the sides of the tray and tape them in place, leaving most of the foil exposed.

This is easy to make. You need a large empty box that held kitchen matches, measuring at least 1" x 2" x 4". The **tray** is the part that slides in and out of the **sleeve.**

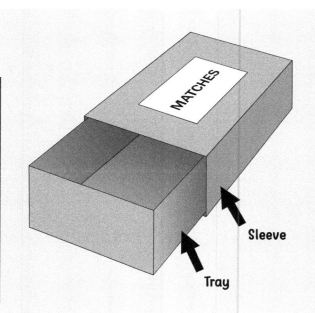

MATCHES

Sleeve

Tray

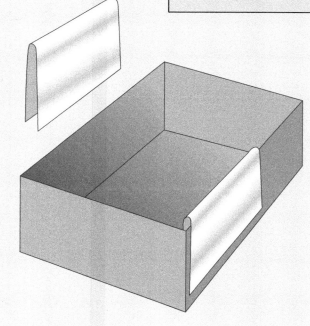

Reuse the complete breadboard setup from Experiment 10. For clarity, the components on the board aren't shown here.

A 9V battery will fit more easily into the tray than your 3-battery holder. Apply power to the circuit and test it by touching the two pieces of foil with finger and thumb.

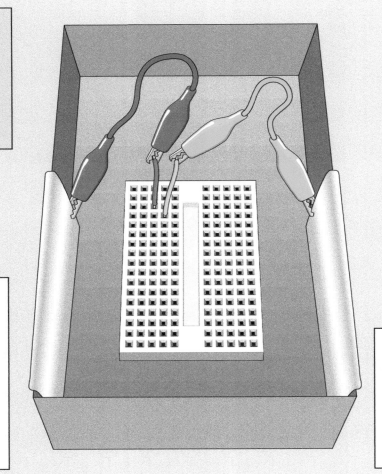

Wrap Scotch tape around the alligator clips to prevent them from touching and creating a short circuit.

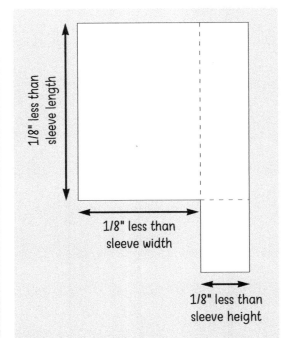

1/8" less than sleeve length

1/8" less than sleeve width

1/8" less than sleeve height

Now cut two pieces of foil to wrap your matchbox sleeve. Trim them about 1/8" smaller than the dimensions of the box.

Wrap the foil over the top and one side of the sleeve. Tape the edges to the sleeve. Fold the foil tab around the end of the sleeve and tape it to the inside.

Use the second piece of foil to wrap the remaining two sides. Do not allow the two sections of foil to touch each other when taping them beside each other.

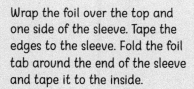

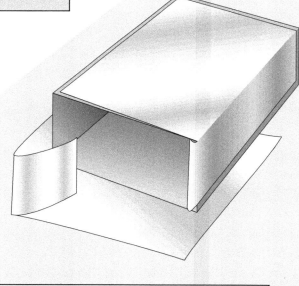

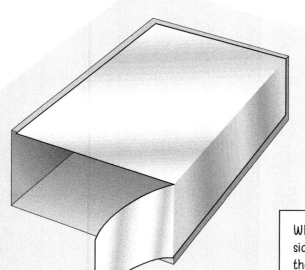

You have to hold the matchbox by the ends, or rest it on the palm of your hand, to avoid activating the beeper when you offer it to someone.

When you slide the tray into the sleeve, the foil on the sides of the tray must connect with the tabs of foil that you wrapped around the ends of the sleeve. When someone holds the sleeve, their skin completes the circuit and the transducer will start squeaking.

Don't forget to disconnect the power after you use this toy. Otherwise, it will run down the battery.

Experiment 12
An Alarming Circuit

A transistor controlled the timer chip in Experiment 10. A phototransistor will work, too.

Now relocate the 10K resistor, and add the phototransistor. Make sure its **short** wire is nearest to the bottom of the board.

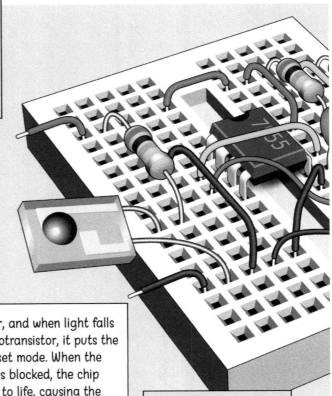

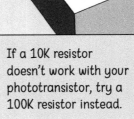

First remove the components that you added in Experiment 10: the transistor, resistor, and jumper wires. Leave all the other components on the board. The left side should look like this.

Apply power, and when light falls on the phototransistor, it puts the chip into reset mode. When the light beam is blocked, the chip comes back to life, causing the transducer to make sound. Position the breadboard so that when an intruder interrupts the light beam, the alarm goes off.

If a 10K resistor doesn't work with your phototransistor, try a 100K resistor instead.

How Did It Work?

In Experiment 10, when the transistor switched on and its resistance went **low,** this woke up the timer. In this experiment, when light to the phototransistor is blocked and its resistance goes **high,** this woke up the timer.

How can the timer chip respond to two opposite signals?

In the schematic, when light falls on the phototransistor, its resistance is low compared with the 10K resistor. So pin 4 of the timer sees a low voltage, and the timer is in reset mode.

When light to the phototransistor is blocked, its resistance goes very high compared with the 10K resistor. Now pin 4 of the timer sees a higher voltage, and the chip wakes up.

Compare the circuit to the schematic in Experiment 10 on page 41, and you'll see I turned the voltage divider upside-down.

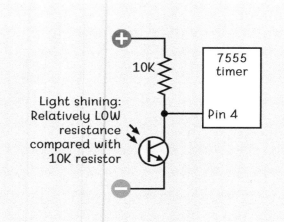

Light shining:
Relatively LOW resistance compared with 10K resistor

10K — 7555 timer — Pin 4

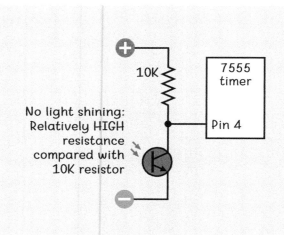

No light shining:
Relatively HIGH resistance compared with 10K resistor

10K — 7555 timer — Pin 4

Alarm Setup

Locate the phototransistor and the laser pen 10 or 20 feet away, and use any two-conductor wire, such as loudspeaker wire, to connect the phototransistor with the breadboard.

To avoid running down your batteries, you can buy a 5VDC AC-DC converter very cheaply online.

Another Idea

Your phototransistor "sees" infrared. Point a TV remote at it, and infrared pulses from the remote will be turned into rapid beeps.

I hope you'll want to know more about electronics. You can move on to my own book, *Make: Electronics*, or many others. I can't guarantee that everything will be as easy as *Easy Electronics*, but I know it will be fun.

Shopping List

The easiest way to obtain parts for the experiments in this book is by ordering a kit. See page 3 for instructions.

If you prefer to shop online, I suggest sites such as **ebay.com**, **mouser.com**, **digikey.com**, or **newark.com**.

Here's a complete list of everything you need.

Batteries. Alkaline AA size. **Quantity: 3.** Note: *Do not use lithium batteries!*

Battery holder for single AA battery. With solder pins or PCB terminals. Eagle Plastic Devices part 12BH311P-GR or similar. **Quantity: 2.**

Battery holder for three AA batteries. With solder pins or PCB terminals. Eagle Plastic Devices part 12BH331P-GR or similar. **Quantity: 1.**

Miniature incandescent light bulb (this is sometimes sold as a "lamp"). The one that I prefer, pictured throughout this book, is rated for 5V and 60mA and has a ceramic base with two short leads. Search for JKL 7361 by JKL components, or CM7361 by Chicago Miniature Lighting, a subsidiary of VCC. If these options are unavailable, a very similar bulb is JKL 7362; it will use more current and will be a bit dim when used with a series resistor, but should work in the experiments. Another option is to use a 6V bulb rated for 40mA or 60mA; these usually have a size E10 screw-thread base, and require a matching socket. Search online for E10 6V bulb, and you will find scientific supply companies selling the bulb and socket for high-school electrical experiments. It may be less bright than an equivalent 5V bulb, but should work. **Quantity: 2.**

Alligator jumper wire (single wire with alligator clip at each end). Any length, but very short ones are more convenient (3" to 6"). **Quantity: 2 red, 2 black, 3 green.**

Slide switch, also known as a **slider switch**. To use it with alligator clips, it should be as large as possible, with pins widely spaced. This can be a problem, as most slide switches today are subminiature. I suggest that the minimum body size is 1/2" or 13mm long, minimum pin spacing 1/8" or 5mm. You want a single-pole, double-throw switch, which may be identified as SPDT, SP2T, 1P2T, or 1PDT. Examples of an acceptable switch are part number PM13B012 by Apem or L102011MS02Q by C&K Components. You will be switching very small currents at only 4.5V, so you do not need to be concerned about maximum voltage or amperage listed for a switch. **Quantity: 1.**

Resistors, quarter-watt, 5% or 10% tolerance. Values 33 ohms, 1K, 10K, and 100K: **2 of each.** Values 2.2K and 3.3K: **1 of each.**

Transistor, 2N3904 NPN bipolar, from any manufacturer. **Quantity: 2.**

LED, low-current type, tinted red. Avago or Broadcom HLMP-D150, or HLMP-D155, or HLMP K-150, or HLMP K-155, for a typical 1.6V forward voltage, 20mA maximum average current but able to respond to 1mA. **Quantity: 2.**

Phototransistor, Lite-On LTR-301 preferred, side-facing NPN type, rated 5V. Alternatively Optek / TT Electronics OP550B. (The O at the beginning of this part number is letter O, the 0 near the end is numeral zero.) The component that you use must be able to pass a constant current of 3mA. If you find that a side-looking phototransistor has both leads of equal length, hold the component with the lens facing you and the leads pointing down, and the right-hand lead is probably the collector (more positive). If in doubt, apply power very briefly. **Quantity: 1.**

Capacitors, electrolytic, rated 10V or higher. 1µF, 10µF, 100µF, and 470µF. **Quantity: 1 of each.**

Capacitors, ceramic. 2.2nF, 10nF, and 100nF. (These values may be written as 0.0022µF, 0.01µF, and 0.1µF.) **Quantity: 1 of each.**

7555 timer chip. Preferred manufacturer is Intersil. If you try using a 555 timer chip, it will consume more current and may not work well at the low voltage in the experiments in this book. **Quantity: 1.**

Mini-breadboard, 17 rows of holes or more. **Quantity: 1.**

Jumper wires, 22-gauge, stripped at both ends, in colors red, green, and black. Length of insulation 1/2": **3 of each color.** Length of insulation 1": **2 of each color.**

Piezoelectric audio transducer with wire leads, DB Unlimited TP244003-1 preferred. Alternatively, CPE-827 from CUI Inc. If you search online, note that "piezo" is often used as the abbreviation for "piezoelectric," and you should search for "piezo speakers" to avoid finding other kinds of transducers, some of which work like microphones. If making a substitution, larger is better (minimum diameter 1" or 25mm). **Quantity: 1.**

Large match box (empty). **Quantity: 1.**

Aluminum foil, piece 6" long.

Scotch tape, any type. Length 18".

Optional Items

Laser pen for triggering phototransistor. **Quantity: 1.**

9V alkaline battery for matchbox toy. **Quantity: 1.**

Magnifying lens (useful for reading part numbers). **Quantity: 1.**

Extra wire (such as **speaker wire**) if you want to build the intrusion alarm with the sensor placed remotely.

Glossary

actuator: The knob, lever, button, toggle, or other movable part on the outside of a switch.

amp: abbreviation for **ampere.**

ampere: The universal unit of electrical **current**. Often abbreviated as **amp**. (An amplifier may also be referred to as an amp.)

base: The central part of a silicon sandwich that controls **current** flowing through a **transistor**.

beeper: A small component that makes a beeping sound, either when a voltage is applied to it, or when electrical pulses are supplied to it. Also known as a **buzzer**, and usually containing a **piezoelectric transducer**.

breadboard: A plastic board perforated with holes into which the **leads** of **components** can be inserted. **Conductors** inside the breadboard create electrical connections between the components.

buzzer: Same as **beeper**.

capacitance: The ability of two conductive parts (such as wires in a **circuit** or **plates** in a **capacitor**) to store electrical energy by maintaining charges of opposite **polarity**.

capacitor: a component that stores electrical energy by holding a **voltage** between two interior **plates**.

ceramic: A type of **capacitor**, often relatively small, in which the **plates** are embedded in ceramic material.

chip: See **silicon chip**.

circuit: Wires and **components** connecting one side of a power source with the other side, so that it can pass **current** through them.

collector: The section of an **NPN transistor** into which **current** flows. (In a **PNP** transistor, the collector is the section out of which current flows.)

component: An electrical part.

contact: The metal element inside a switch that creates a connection with a second contact (or multiple contacts) when the **actua-tor** of the switch is moved or pressed.

conductor: A substance, usually made of metal, that has very low **resistance**.

current: The flow of electrical charge (composed of **electrons**) through a wire or **component**, during a fixed period of time.

electrolytic: A common type of **capacitor**.

electron: A subatomic particle that revolves around the nucleus of an atom and carries an electrical charge. A flow of electricity consists of electrons.

emitter: The section of an **NPN transistor** out of which **current** flows. (In a **PNP** transistor, the emitter is the section into which current flows.)

farad: The universal unit of **capacitance**.

frequency: The number of times that a state such as **voltage** fluctuates per second.

hertz (Hz): A unit that measures **frequency**.

incandescence: The emission of light by a

substance as a function of it being hot.

integrated circuit: A very small **circuit**, including **components**, etched onto a **chip** of silicon. See **silicon chip**.

kilohertz: 1,000 **Hertz**, abbreviated kHz.

kilohm: 1,000 **ohms**, abbreviated with letter K.

lamp: Sometimes used to mean a miniature **incandescent** light bulb.

lead: A lead (pronounced "leed") is usually a wire sticking out of a **component**.

megohm: 1,000,000 **ohms**.

microfarad: One millionth of a **farad**, often represented by µF, or uF if the character to represent Greek letter mu is not available.

milliamp: Abbreviation for **milliampere**, which is one-thousandth of an **ampere**. The abbreviation mA is often used.

millivolt: One-thousandth of a **volt**. The abbreviation mV is often used.

nanofarad: One-thousandth of a **microfarad**. The abbreviation nF is often used.

NPN: Acronym for negative-positive-negative, referring to relative **polarities** of layers of silicon inside a **transistor**.

ohm: The universal unit that measures electrical **resistance**.

picofarad: One-thousandth of a **nanofarad**. The abbreviation pF is often used.

plate: The metal part inside a **capacitor** that stores **voltage** as a charge opposite to that on a second plate.

phototransistor: A **transistor** in which the **base** responds to light instead of to electric **current**.

piezoelectric effect: The effect of producing **voltage** by stressing a crystal, or creating stress on a crystal by applying voltage. The abbreviation **piezo** is often used. In a piezoelectric **transducer**, fluctuating voltage causes a thin crystal wafer to vibrate, creating waves of air pressure perceived as sound. The piezo transducer may also function as a microphone, converting sound into fluctuations of voltage.

PNP: Acronym for positive-negative-positive. See **NPN**.

polarity: The condition of a component or a circuit where one part has a higher voltage, or must have a higher voltage, than another part.

resistance: The ability of a **conductor**, a **resistor**, or some other **component** to restrict **current** and **voltage**. Usually measured in **ohms**.

semiconductor: A solid substance, often silicon, that is used in a **component** such as a **transistor** to control the flow of electricity.

silicon chip: A wafer of silicon into which has been etched an **integrated circuit**. The chip is sealed into a small package, usually made of plastic. The package itself is also referred to as a chip.

transducer: A **component** that converts physical force into electricity, or vice-versa. See **piezoelectric effect**.

transistor: a **semiconductor** that will act as a switch that turns **current** on and off, or as an amplifier that multiplies the current, depending on current flowing into the **base**.

volt: The universal unit of electrical pressure. See **voltage**.

voltage: The electrical charge in one part of a **circuit** or **component**, relative to another part. A greater concentration of **electrons** creates a higher voltage. This is measured in **volts**.

Charles Platt

VOL. 1

Encyclopedia of
Electronic
Compon...

Power Sources & C...
Resistors · Capacitors · Indu...
Switches · Encoders · Relays...

O'REILLY

Charles Platt with Fredrik Jansson

VOL. 2

Encyclopedia of
Electronic
Compon...

Signal Processing
LEDs · LCDs · Audio · Thyrist...
Digital Logic · Amplification...

Make:

VOL. 3

Charles Platt and Fredrik Jansson

Encyclopedia of
Electronic
Components

Sensors

Location · Presence · Proximity · Orientation ·
Oscillation · Force · Load · Human Input · Liquid and
Gas Properties · Light · Heat · Sound · Electricity

Vol. 1: ISBN: 978-1-449-33389-8 US $24.99

Vol. 2: ISBN: 978-1-4493-3418-5 US $29.99

Vol. 3: ISBN: 978-1-4493-3431-4 US $29.99

All Your Questions Answered

Whenever you want to know—what can this component do? How should I use it? What other components could I use instead?

In three volumes, the *Encyclopedia of Electronic Components* provides quick, authoritative guidance.

Comprehensive and clearly written, with copious illustrations, this is the only reference work of its kind.

Learning Notes

Name _____